COMPETITIVE CHEMISTRY 1

INTRODUCTION

This objective chemistry series provides a basic and challenging problem of chemistry from particular topics. It can be used to brush up ones basics and checking up the preparation level of particular topic. It is equally helpful to the traditional classes as well as competitions. It can be also taken as a revision material for any competition which includes the test of basic chemistry. If you want to grasp the subject before practicing these multiple choice questions, you can go through the website http://www.ncert.nic.in/ncerts/textbook/textbook.htm and down load the free copy of science books and after having command on the topic practice it.

If you have any query or suggestion about this series you can send your suggestion at uk2594@gmail.com.

CONTENTS

1. MATTER IN OUR SURROUNDINGS

SOME IMPORTANT POINTS

➤ The things that occupy space and have mass are known as matter.

➤ A matter is made up of very small particles.

➤ These p[articles of matter have spaces between them, these are continuously moving, these attract each other.

➤ The intermixing of these particles in gaseous phase is known as diffusion.

➤ Simply matter is found in three states: solid, liquid and gases.

➤ The force of attraction is maximum in solids and least in gases.

➤ The intermolecular space is maximum in gases and least in solids.

➤ The states of matter can be changed by changing the temperature or pressure.

➤ In sublimation solid directly changes into gases.

➤ The phenomenon of changing of a liquid into its vapour at any temperature below its boiling point is known as evaporation.

➤ The rate of evaporation depends upon the surface area, temperature, humidity and the speed of wind.

➤ Evaporation causes cooling.

1. MATTER IN OUR SURROUNDINGS

1. Everything in this universe is made up of?

    a) Metal   b) Non metal   c) Matter   d) None of these

2. What is the example of matter?

    a) Star   b) Plants   c) Clouds   d) All of these

3. Five basic elements are?

    a) air, Earth, fire, sky and water   b) plant, air, sky, soil &Matter

    c) fire, material, earth, land   d) None of these

4. How Many types of classification of matter are?

    a) Two   b) three   c) four   d) five

5. Matter is made up of

    a) soil   b) rock   c) Particles   d) none of these

6. What is the colour of water?

    a) Red   b) Sky colour   c) colourless   d) golden

7. The particles of matter are?

    a) small   b) long   c) very small   d) None of these

8. Space between particles of one type of matter are?

    a) big space b) small space   c) enough space d) not space

9. Particles of Matter are -------------- moving.

    a) Straight line   b) continuously   c) both   d) None of these

10. As the temperature rises, particles move?

    a) slowly-slowly   b) faster   c) None of these

11. The kinetic energy of the particles …………………. On increasing temperature.

a) decreases   b) move backward c) increases d) None of these

12.   Intermixing of particles of two different type matters on their own is called?

a) Collision   b) Matter   c) Particles   d) diffusion

13.   Particles of matter attract:

a) each other   b) together   c) other object   d) None of these

14.   Which states of matter arise due to the variation in the characteristics of the particles of matter?

a) solid   b) Liquid   c) gas   d) all of these

15.   What are the example of solid, liquid, gas:

a) rock, water oxygen   b) soil, tea, coffee

c) H_2, H_2SO_4, Nitrogen   d) None of these

16.   Solids have a tendency to maintain their:

a) Shape   b) Volume   c) Velocity   d) Both a and b

17.   Liquids have no fixed shape but have a:

a) fixe volume   b) Fixed density   c) None of these

18.   Liquids flow and change shape so they are not rigid but can be called?

a) diffusion   b) fluid   c) matter   d) None of these

19.   Solids and liquids can diffuse into:

a) solids   b) liquid   c) gaseous   d) None of these

20.   the gases from the atmosphere diffuse and dissolve in?

a) Water   b) salt   c) snow   d) None of these

21.   Solids, liquids and gases can diffuse into?

a) liquids   b) solids   c) gases   d) all of these

22. The rate of diffusion of liquid is higher than that of?

   a) gases   b) solids   c) none of these   d) a & b

23. Gases are highly compressible as compared to?

   a) solids and liquids   b) solids and gases

   c) liquies and gases   d) None of these

24. CNG is used as?

   a) Fuel   b) Breathe   c) Both a and b   d) None of these

25. The aquatic animals can breathe under water due to the presence of dissolve oxygen in?

   a) Earth   b) under the Earth   c) Water   d) None of these

26. Define density?

   a) The mass per unit volume of a substance

   b) The mass per unit volume square of a substance

   c) Both a and b

   d) None of these

27. On increasing the temperature of solids which energy of the particles increases

   a) Kinetic energy   b) Potential energy

   c) Mechnical energy   d) None of these

28. Due to the increase in kinetic energy the particles stars:

   a) Moving   b) vibrating   c) None of these

29. The energy supplied by heat overcomes the force of attraction between the

   a) Matter   b) object   c) particles   d) all of these

30. The temperature at which a solid melts to become a liquid at the atmospheric pressure is called its?

a) individual point b) point   c) centre   d) melting point

31. The melting point of ice is?

a) 273.16Kg   b) 273.16J   c) 273.16N   d) 273.16K

32. The Process of melting, which is change of solid state into liquid state is also known as?

a) fussion   b) fusion   c) collision   d) None of these

33. When we supply heat energy to water particles, particles start moving even:

a) faster   b) slowly   c) None of these

34. The temperature at which a liquid starts boiling at the atmospheric pressure is called?

a) boiling point   b) melting Point   c) latent heat
d) None of these

35. Boilling is as Phenomenon of?

a) Albert   b) Bulk   c) microscopic   d) none of these

36. Particles from the bulk of the liquid gain enough energy to change into the?

a) Liquid state   b) solid state   c) gaseous state   d) vapour state

37. The state of matter can be changed into another state by changing the?

a) volume   b) Mass   c) temperature   d) None of these

38. What is the SI unit of temperature?

a) K   b) J   c) Kwh   d) None of these

39. The SI unite of volume is?

a) m^2 b) m^3 c) m/s   d) metre

40. What is the common unit of volume is

a) m^3 b) litre   c) metre   d) m/s

41. What is unit of pressure?

    a) pascal   b) Kwh   c) Kgm/s^2 d) None of these

42. The Pressure of air in atmosphere is called?

    a) atmospheric Pressure b) Normal Pressure

    c) Pressure   d) None of these

43. A change of state directly from solid to gas without changing into liquid is called?

    a) sublimation   b) vaporization   c) vibration d) None of these

44. The solid carbon dioxide is also known as?

    a) dry ice   b) fusion   c) None of these

45. The change of a liquid into vapors at any temperature below its boiling point is called?

    a) Evaporation   b) fusion   c) collision   d) None of these

46. The state of matter can be changed by changing temperature or?

    a) Volume   b) pressure   c) velocity   d) None of these

47. What is the unit of weight?

    a) K   b) Newton   c) Kg   d) Pa

48. What is the unit of length?

    a) meter   b) meter/second   c) m/s^2 d) None of these

49. The particles gain energy from your palm or surrounds and evaporate causing the palm to feel?

    a) Cool   b) hot   c) both   d) None of these

50. During summer we perspire more because of the mechanism of our body which keeps us?

    a) Cool   b) hot   c) None of these

51. The sun and the stars glow because of the presence of:

    a) Plasma   b) heat   c) light   d) None of these

52. Who had done some calculations for a fifth state of matter?

    a) S.N. Bose   b) Isaac Newton   c) None of these

53. When Indian physicist S.N. Bose had done some calculations for a fifth state of matter

    a) 1930   b) 1920   c) 1940   d) 1950

Answers:

Ques.	Ans.	Ques.		Ques.		Ques.		Ques.		Ques.	
1	C	11	C	21	A	31	D	41	A	51	A
2	D	12	D	22	B	32	B	42	A	52	A
3	A	13	A	23	B	33	A	43	A	53	B
4	A	14	D	24	A	34	A	44	A		
5	C	15	A	25	C	35	B	45	A		
6	C	16	D	26	A	36	D	46	B		
7	C	17	A	27	A	37	C	47	B		
8	C	18	B	28	B	38	A	48	A		
9	B	19	B	29	C	39	B	49	A		
10	B	20	A	30	D	40	B	50	A		

2. IS MATTER AROUND US PURE

SOME IMPORTANT POINTS

- Mixture is a combination of two or more substances in any proportion.
- Mixtures components can be separated using appropriate technique.
- In homogeneous mixture each and every part of mixture shows uniform composition while it is non uniform in non homogeneous.
- Solution is a homogeneous mixture. The major components of solution are known as solvent while minor is termed as solute.
- Colloids and suspensions are heterogeneous mixture of substances. In suspension the solute particles are seen by naked eye. In colloids the solute particles are not seen by naked eye but can scatter the light.
- We can separate the dye from ink evaporation.
- Configuration is useful for separating cream from milk.
- Two immiscible liquids are separated using separating funnel.
- Mixture of Ammonium and salt can be separated by sublimation.
- Mixture of two miscible liquids can be separated by fractional distillation method.
- Air is a homo9geneous mixture of gases and these components can be separated from air.
- Crystallization technique is useful to obtain pure copper sulphate from impure sample.
- During the burning of candle both chemical and physical changes occur.
- Elements and compounds are the pure substances. Properties ,of compounds are different from its constituents.

2. IS MATTER AROUND US PURE

1.   What is a mixture?

   a. Mixtures is constituted by one kind of pure matter.

   b. Mixture is constituted by more than one kind of pure or impure matter.

   c. Both a and b

   d. None of these

2.   From which process dissolved sodium chloride can be separated from water?

   a. Centrifugation   b. Filtration   c. Evaporation   d. Distillation

3.   Sugar is a?

   a. Pure matter   b. Impure Matter   c. Mixure   d. None of these

4.   What is a solution?

   a. a heterogeneous mixture of two or more substances.

   b. a pure form of matter

   c. Homogeneous mixture of two or more substances

   d. None of these

5.   Which of the following is a solution?

   a. sugar   b. Kerosene oil   c. water   d. Soda water

6.   Which of the following is not a solution?

   a. Soda water   b. Lemonade   c. Air   d. None of these

7.   What is alloy?

   a. mixture of two or more metals or a metal and non metal

b. mixture of two or more metals

c. Mixure of two or more non metals

d. None of these

8.   Which cannot be separated by physical method?

   a. Solder   b. Bronze   c. Brass   d. All of these

9.   Brass has the % of zinc and copper respectively:

   a. 70, 30   b. 50, 50   c. 30, 70   d. None of these

10.   In a solution the component present in larger amount is known as?

   a. Solute   b. Solvent   c. Solution   d. All of these

11.   In a Solution the compound present in lesser quantity is known as?

   a. Solute   b. Solvent   c. Solution   d. All of these

12.   A solution of iodine in alcohol is known as ?

   a. Compound iodine   b. Tincture of iodine

   c. Pure form of iodine   d. None of these

13.   Air is a homogenous mixture of a number of gases .its two main constituents are ?

   a. Oxygen (21%) and Nitrogen (78%)   b. Oxygen (20%) and Nitrogen (78%)

   c. Oxygen (22%) and Nitrogen (78%)   d. Oxygen (21%) and Nitrogen (77%)

14.   A solution is _____ mixture .

   a. Heterogeneous mixture   b.Colloidal mixture

   c. Suspension mixture   d. Homogeneous mixture

15.   1nm =?

   a. 10^{-7}m   b. 10^{-9}m   c. 10^{-8} m   d. 10^{-10}m

16. Which can scatter beam of light?

    a. Colloidal   b. Solution   c. Suspension   d. All of these

17. In which solute particle settle down when left undisturbed?

    a. Solution   b. Suspension   c. Colloidal   d. Both (b) &(c)

18. The amount of the solute present in the saturated solution at particular temperature is called its _____ .

    a. Solubility b. Solvity   c. Colloidility   d. None of these

19. Concentrated of solution = ?

    a. Amount of solute /Amount of solution   b. Amount of solute/ Amount of solvent

    c. Both (a)&(b)   d. Amount of solvent /Amount of solute

20. Mass by mass percentage of a solution =?

    a. Mass of solution /Mass of solute *100   b. Mass of solution /Mass of solute *1000

    c. Mass of solute /Mass of solution*1000   d. Mass of solute /Mass of solution*100

21. Mass by volume percentage of solution =?

    a. Mass of solution /Volume of solution*100

    b. Mass of solution/volume of solute*100

    c.Mass of solute /volume of solution*100

    d. None of these

22. A solution contains 30g of sugar in 430 g of water .Calculate the concentration in terms of mass by mass percentage of the solution ?

    a. 6.78%   b. 6.52%   c. 7.52%   d. None of these

23. In which of the following we can see the particle of substance to the naked eye?

a. Suspension   b. Colloidal   c. Solution   d. None of these

24.   Which of following is a heterogeneous?

a. Suspension   b. Colloidal   c. Both (a)&(b)   d.None of these

25.   The scattering of a beam of light is called _____ .

a. Tincture effect   b. Tyndall effect   c. Scattering   d. Natural effect

26.   From which technique we can separate the colloidal particle ?

a. Sublimation   b. Evaporartion   c.Filtration   d. Configuration

27.   In which of the following is a colloidal ?

a. Cheese   b. Chips   c. Ammonium chloride   d. None of these

28.   In which of the following is a type of colloidal?

a. Aerosol   b. Emulsion   c. Sol   d. Gel

29.   When dispersed phase is solid and dispersing medium is liquid the type of colloidal

is?

a. Solid sol   b. Emulsion   c. Sol   d. Foam

30.   Shaving cream is example which type of colloids?

a.Aerosol   b. Sol   c. Solid Sol   d. Foam

31.   How can we obtain coloured component (DYE) from Blue /Black ink ?

a. Evaporation   b.Sublimation   c. Both (a)&(b)   d. Filtration

32.   How can we separate cream from milk?

a. Evaporation   b. Centrifugation   c. Separating funnel   d. None of these

33.   Which of the following method used in diagnostic laboratories for blood and urine

test ?

a. Centrifugation   b. Evaporation   c. Filtration   d. All of these

34. How can be we separate a mixture of two immiscible liquids ?

a. Evaporation   b. Sublimation   c. Filtration   d. None of these

35. In the extraction of iron from its ore which technique is used?

a. Separating funnel   b. Centrifugation   c. Sublimation   d. All of these

36. Which is the following is a example of solids which sublime ?

a. Ammonium chloride   b.Camphor   c. Antracene   d. All of these

37. How we can separate a mixture of salt and camphor ?

a. Sublimation   b.Evaporation   c. Chromatograph   d. Filtration

38. Separation of dyes in black ink using ?

a. Chromatography   b. Evaporation   c. Sublimation   d. Separating funnel

39. To separate drugs from blood which method is used?

a. Centrifugation   b.Chromatography   c. Filtration   d. separating funnel

40. Which method is used for the separation of components of a mixture containing

two miscible liquid?

a. Distillation   b. Chromatography   c. Sublimation   d. None of these

41. How we can obtain pure copper sulphate from an impure sample?

a. Chromatography   b. Crystallisation   c. Sublimation   d. Distillation

42. The method used in separation of crystals of alum from impure samples?

a. Chromatography   b. Crystallisation   c. Sublimation   d. Distillation

43. Who was the first scientist to use the term element in 1661?

a. Antoine Laurent   b. Robert boyle   c. Antoine Lavoiser   d. None of these

44. Which of the following is a properties of metal ?

    a. Malleable   b. Ductile   c. Sonours   d. All of these

45. Which of the following a metal ?

    a. Copper   b. Bromine   c. Iodine   d. Oxygen

46. Which is following a non metal?

    a. Gold   b. Platinum   c. Silver   d. Carbon

47. Which is a metalloid?

    a.Polonium   b. Germanium   c. Boron   d. All of these

48. Which two elements are liquid?

    a. Mercury , lead   b. Mercury , Bromine   c. Bromine , Lead   d. Mercury, Oxygen

49. Which a compound out of a following?

    a. Water   b. Copper   c. Stainless steel   d.Solder

50. Which of the following is physical change?

    a. Burning of alcohol   b. Formation ice from water

    c. Both (a) & (b)   d. None of these

Answers:

Q	A	Q	A	Q	A	Q	A	Q	A
1	B	11	A	21	C	31	A	41	B
2	C	12	B	22	B	32	B	42	B
3	A	13	A	23	A	33	A	43	B
4	C	14	D	24	C	34	D	44	D
5	D	15	B	25	B	35	A	45	A
6	D	16	A	26	D	36	D	46	D
7	A	17	B	27	A	37	A	47	D
8	D	18	A	28	D	38	A	48	B
9	C	19	C	29	C	39	B	49	A
10	B	20	D	30	D	40	A	50	B

3. ATOMS AND MOLECULES

SOME IMPORTANT POINTS

- ➢ Law of conservation of mass states that mass can neither be created nor be destroyed in a chemical reaction.
- ➢ In a pure substance the amount of elements present in a fixed proportion by mass. This is known as the law of constant proportion or the laws of definite proportion.
- ➢ The atomic theory was given by John Dalton. It tells that the atom is the smallest particle present in matter. Atoms are indivisible. Atoms of different elements are of different masses and different properties.
- ➢ A group of two or more atoms that are chemically bonded together are known as mo0lecules.
- ➢ The combining capacity of an atom is known as valency.
- ➢ The Avogadro constant represents the number of $6.022*10^{23}$.
- ➢ Mass (in gram) of 1 mol of a substance is called to molar mass.

3. ATOM AND MOLECULES

1.   What is the meaning of atom?

a. Derive   b. Indivisible   c. Divisible   d. Small

2.   Who gave the laws of chemical combination?

a. Dalton   b. L.Lavoisier   c. Joseph L. Proust   d. Both (a) & (b)

3.   In water, the ratio of mass of hydrogen to the mass of oxygen is?

a. 1:9   b. 14:3   c.1:8   d. 8:1

4.   In ammonia the ratio of nitrogen and hydrogen in the ratio?

a. 3:14   b. 14:3   c.13:4   d. 13:5

5.   Who stated the law of constant proportion?

a. L.Lavoisier   b. Joseph L.Proust   c. Dalton   d. Both (a) & (b)

6.   Who give the basic theory about the nature of matter?

a. E.Goldstien   b. J.J Thompson   c. John Dalton   d. Albert Einstien

7.   Dalton theory based on?

a. The law of conservation of mass.   b. The law of constant proportion

c. The laws of chemical combination   d. The law of definite proportion

8.   Atomic radius is measured in ?

a. Picometer   b. Nano meter   c. Mano meter   d. None of these

9.   How many nano meters in 1 meter?

a. $1/10^9$ b. 10^{9nm} c.10^{-9nm} d. None of these

10.   Our entire world is made up of?

a. Cell   b. Atom   c.Nucleus   d. All of these

11.   The symbol of cobalt is?

        a. Cu   b. Co   c. Co   d. C

12.   The symbol of lead is?

        a. PB   b. Pb   c. pb   d. Pa

13.   The symbol of tungsten?

        a. Tu   b. Te   c. Cs   d. W

14.   The symbol of Au is used for ?

        a. Silver   b. Arsenic   c. Gold   d. Argon

15.   The atomic no. Of tungsten is?

        a. 53   b. 83   c.74   d. 89

16.   The atomic no. Of iodine is?

        a. 49   b. 69   c.53   d. 43

17.   Helium is a?

        a. Metaloid   b. Metal   c.Non metal   d. All of these

18.   Earlier we defined the atomic mass unit as _____.

        a. u   b. v   c.amu   d. Am

19.   In carbon monoxide, carbon and oxygen combined in the ratio of?

        a. 1:8   b. 3:4   c.4:3   d. 6:8

20.   Who gave the atomic theory?

        a. E. Goldstien   b. John Wolfgang   c. John. Dalton   d. Both (a) & (b)

21. In these of the following which is very reactive?

   a. He   b. Xe   c. As   d. O

22. How many isotopes of carbon exist in nature?

   a. 5   b. 2   c.12   d.15

23. The atomic mass of the chlorine is?

   a. 35.9   b. 34   c.35.7   d. 35.5

24. How many isotopes of hydrogen exist in?

   a. 0   b. 3   c.4   d. None of these

25. The group of two or more atoms is known as?

   a. Element   b. Molecule   c. Compound   d. Mixture

26. The atom of oxygen is known as ?

   a. Monatomic molecule   b. Tetra atomic molecule

   c. Diatomic molecule   d.Triatomic molecule

27. The formula of ozone is?

   a. O_2 b. OZ   c.O_3 d. O

28. The no. Of atoms consisting a molecule is known as its_____

   a. Monoatomic   b. Atom city   c.Molecity   d. Atomic mass

29. Carbon is a?

   a. Metal   b. Non metal   c. Metaloid   d. None of these

30. The ratio of masses of H:O is water is?

   a. 1:8   b. 1:2   c. 8:1   d. 3:8

31. Carbon is a?

 a. Cation   b. Anion   c. Both (a) & (b)   d. None of these

32. What is the chemical formula of ammonium?

 a. $HNCL_3$ b. $NHCL_4$ c. NH_3CL d. NH_4CL

33. What is the formula of ammonium sulphate?

 a. $(NH_4)SO_4$ b. NH_4SO_4 c. $(NH_4)_2SO_4$ d. $(NH)_4 SO_{12}$

34. What is the name of compound represent by formula $Al_2(SO_4)_3$?

 a. Aluminium Sulphur   b. Argon sulphur

 c. Aluminium Sulphate   d. AluminiumSulphide

35. Formula of sodium nitrate?

 a. NaNo   b. $NANO_3$ c. $NaNo_3$ d. $NaNo_3$

36. What is the molecular mass of ammoia?

 a. 16u   b. 13u   c. 8u   d. 18u

37. What is the molecular mass of ammonium chloride?

 a. 35.5u   b. 53.5u   c. 55.5u   d. 63.5u

38. What is the molecular mass of potassium nitrate?

 a. 100u   b. 99.2u   c. 101.1u   d. 101.3u

39. The atomic mass of Argon is?

 a. 29.98u   b. 39.9u   c. 39.8u   d. 39u

40. Calculate the formula unit mass of K_2CO_3?

 a. 128u   b. 138.3u   c. 138u   d. 138.2u

41. Molecular mass of CH_3OH?

a.31.22u   b.33.34u   c.32u   d.32.2

42.   Avogadro constant is representing?

        a.N_1 b.AC   c.N_o d.AN_o

43.   What is the value of Avogadro constant is ?

        a.$6.673*10^{-11}$ b. $6.667*10^{23}$ c. $6.023*10^{23}$ d. $6.022*10^{23}$

44.   How many dozen in 1 gross?

        a. 1   b.13   c.6   d.12

45.   Gram atomic mass is known as?

        a. Atomic mass   b. Atomicity   c. Molar mass   d. Mass number

46.   Calculate the no. of moles in 13g of hydrogen atom?

        a.12   b.13   c.11   d.14

47.   Calculate the mass of 0.5 mole of N_2 gas?

        a.7g   b.14g   c.12g   d.6g

48.   Calculate the no. of atom present in 2 moles?

        a.$6.022*10^{23}$ b. $12.044*10^{23}$ c.$6.023*10^{23}$ d. $12.022*10^{23}$

49.   Calculate the no. of atoms present in 12g of hydrogen atom?

        a. $72.022*10^{23}$ b.$72.264*10^{23}$ c.$6.022*12^{23}$ d.$72.04*10^{23}$

50.   Calculate the no. of atoms present in 69g of sodium atom?

        a.$6.022*10^{23}$ b.$18.066*10^{22}$ c.$19*10^{23}$ d.None of these

Answers:

Q	A	Q	A	Q	A	Q	A	Q	A
1	B	11	C	21	D	31	B	41	C
2	D	12	B	22	B	32	D	42	C
3	C	13	D	23	D	33	C	43	D
4	B	14	C	24	B	34	C	44	D
5	B	15	C	25	B	35	C	45	C
6	C	16	C	26	C	36	D	46	B
7	C	17	C	27	C	37	B	47	B
8	B	18	C	28	B	38	C	48	B
9	B	19	C	29	B	39	B	49	B
10	B	20	C	30	B	40	D	50	D

4. STRUCTURE OF ATOM

SOME IMPORTANT POINTS

- The electron was identified by J.J Thomson and the proton was discovered by E. Goldstein.
- An atom has sub atomic particles, hence the atom is divisible.
- The charge on proton is equal in magnitude to the charge on electron but opposite in sign.
- The atomic nucleus was discovered by the Rutherford.
- The neutron was discovered by James Chadwick and it has no charge.
- The electron has negative charge, proton positive and neutron charge less.
- The mass of electron is about 1/2000 times of proton and the mass of proton and neutron is taken as 1 unit.
- The atomic number is the number of proton or electron in a neutral atom.
- Mass number is the number of proton and neutrons present in the nucleus of an atom.
- Isotopes are the atoms of same element which have same atomic number but different atomic mass.
- The first model for the structure of an atom was given by J.J Thomson.

4. STRUCTURE OF ATOM

1.   The presence of new radiation in a gas discharge detected by J.J Thomson was?

      a. Anode   b. Cathode   c. Canal rays   d. All of above

2.   Electrically charged object:

      a. Attract oppositely charged object   b. Attract uncharged object

      c. Repel uncharged object   d. None of these

3.   Which one is not subatomic particle?

      a. Neutron   b. Neuron   c. Electron   d. All of above

4.   Which one is negative charged atomic part?

      a. Proton   b. Neutron   c. Electron   d. All of above

5.   The mass of a proton is taken as one unit and its charge as:

      a. Plus   b. minus   c. none of these

6.   An electron is represented as:

      a. e^+ b. P^+ c. e^- d. P^-

7.   The three main subatomic particles are?

      a. Neutron   b. Proton   c. Electron   d. All of these

8.   The fundamental atomic particle having one unit positive charge is?

      a. Proton   b. Neutron   c. Electron   d. None of these

9.   The ray's travels in straight line called?

      a. Cathode rays   b. X rays   c. Anode rays   d. None of these

10.   _____ was the first one to propose model for the structure of an atom?

      a. Gaelili Galilio   b. J.J Thompson   c. Mark Henry   d. Albert Einstein

11. Cathode rays produce _____ when the strike against the surface of hard Metals?

a. X rays   b. Anode rays   c. Cathode rays   d. All of Above

12. Anode rays carry?

a. Positive charge   b. Negative charge   c. Neutral   d. None of these

13. In the model of J.J Thompson the rate of seeds in watermelon was as?

a. Proton   b. Electron   c. Neutron   d. None of these

14. An atom was a sphere of positive electrically in which were embedded numbers of electron sufficient to neutralize the positive charge?

a. Atom model of J.J Thompson   b. Atom model of Neils Bhor

c. Atom model of Rutherford   d. None of these

15. In the Thompsons model of atom, the sphere is positive charge is called ?

a. Proton   b. Electron   c. Neutron   d. All of above

16. The small heavy positively charged body present within the atom was called?

a. Neuron   b. Nucleus   c. Electron   d. Nucleus

17. Which type of particle Rutherford use in his experiment?

a. . α particle   b. B particle   c. γ particle   d. Atomic particle

18. During the experiment of Rutherford we used _____

a. Silver foil   b.Gold foil   c. Copper foil   d. Litmus paper

19. Rutherford's experiment α particle are doubly charged ?

a. Hydrogen ion   b. Helium ion   c. Argon ion   d. None of these

20. In the Rutherford α particle scattering gave totally unexpected result?

a. Most of α particle passed straight line through the gold foil

b. Some of α particle deflected in a small angle

c. Some of the α destroyed

d. Both (a) & (b)

21. According to Rutherford if your fire a 15 mg shells at a piece of tissue paper and it?

   a. Comes back and hit you   b. Go forward   c. Burn   d. None of these

22. The father of nuclear physics is?

   a. Stefen Howkins   b. E.Rutherford   c. Neils Bhor   d. Isaac Newton

23. According to Rutherford the size of nucleus is very small as compared to the size of?

   a. Nucleus   b. Atom   c. K Shell   d . α particle

24. Which order of shell is correct?

    a. M L K N   b. K L M N   c. L K M N   d. M N K L

25. The subatomic particle which revolves the nucleus?

   a. Protein   b. Neutron   c. Electron   d. All of above

26. There is a gap in the atoms which is causes of?

   a. Passing of α particle   b. Deflection of α particle in same path

   c. α particle deflection in 180   d. All of above

27. The radius of the nucleus is about …….. times less than the radius about the atom?

   a. 10^{11} b. 10^{5} c. 10^{12} d. 10^{9}

28. There is a positive charged centre in an atom called?

   a.Electron   b. Nucleus   c. orbit   d. None of these

29.   Which sub- atomic particle is Neutral?

        a. Proton   b. Electron   c. Neutron   d. All of Above

30.   Which of the level shell are first and last?

        a. K&N   b. LM   c. LK   d. MN

31.   The number of an element is equal to the number........ in the element ?

        a. Electron   b. Proton   c. Neutron Proton   d.Neutron

32.   Pb is symbol of?

        a. Element   b. compound   c. mixture   d. None of these

33.   Calculate the number of electron, Proton and Neutrons in sodium atom .

        Given that atomic number of sodium is 11 and mass number is 23?

        a. 20   b. 21   c.13   d. 12

34.   Number of neutron is equal to?

        a. Mass number – Atomic number   b. Atomic mass - Atomic number

        c. Number of proton =Atomic number   d. None of these

35.   Rutherford's scattering experiment showed for the first time that the atom has?

        a. Electron   b. Proton   c. Nucleus   d. Neutrons

36.   Proton was discovered by?

        a. J.J Thompson   b. Rutherford   c. Chadwick   d. Goldstein

37.   The atomic number of helium is?

        a. 2   b. 1   c. 4   d. 13

38.   The maximum number of shell present in the nth shell is equal to?

        a. $n2^x$ b. $2n^2$ c. $2n^{x2}$ d. Xn^2

39. The electronic configuration of sodium is?

   a. K2 L8 M7   b. K2 L8 M1   c. K2 L8 M3   d. K2 L8 M2

40. The electron revolves the nucleus on its?

   a. Orbit   b. Triangle path   c. straight   d. None of these

41. The electron present in the outermost shell of the atom of an element is

   called?

   a. Valence shell   b. Valence electron c. Both (a) & (b) d. None of these

42. The electronic configuration of chlorine is?

   a. K2 L8 M7   b. K2 L8 M5   c. K2 L8 M3   d. K2 L8 M6

43. The some element which have same atomic number but different mass number

   are called ?

   a. Isobar   b. Isotopes   c. Both (a) & (b)   d. None of these

44. The element which have some chemical property & different physical property?

   a. Isotopes   b. Isobar   c. Both (a)&(b)   d. None of these

45. The different elements which have different atomic number but same mass

   Number called?

   a. Isotopes   b. Isobar   c. Both (a)&(b)   d. None of these

46. Element with negative valency are?

    a. Always metals   b. Always metalloids

   c. Either metals or non metals   d. Always non metals

47. The first model of an atom was given by?

   a. N.Bohr   b. E.Goldstien   c. J.J Thompson   d. Rutherford

Answers:

Q	A	Q	A	Q	A	Q	A	Q	A
1	A	11	A	21	A	31	B	41	A
2	A	12	A	22	B	32	A	42	A
3	B	13	B	23	B	33	D	43	B
4	C	14	A	24	B	34	A	44	A
5	A	15	A	25	C	35	C	45	B
6	C	16	B	26	D	36	C	46	D
7	D	17	A	27	B	37	A	47	C
8	A	18	B	28	B	38	B		
9	C	19	B	29	C	39	B		
10	B	20	D	30	A	40	A		

NOTES

www.ingramcontent.com/pod-product-compliance
Lightning Source LLC
Chambersburg PA
CBHW080624180526

45168CB00007B/3046